LIFE-SIZE
guide to
NEW ZEALAND
BIRDS

Rod Morris

RANDOM HOUSE
NEW ZEALAND

National Library of New Zealand
Cataloguing-in-Publication Data

Morris, Rod, 1951-
Life-size guide to New Zealand birds / Rod Morris.
Includes index.
ISBN 1-86941-527-2
1. Birds—New Zealand—Pictorial works. 2. Birds—New Zealand—
Identification. I. Title.
598.0993022—dc 21

A RANDOM HOUSE BOOK
Published by
Random House New Zealand
18 Poland Road, Glenfield, Auckland, New Zealand
www.randomhouse.co.nz

First published 2002

ISBN 1 86941 527 2

Design and Illustration: Matthew Trbuhovic, The Bureau Interactive Ltd
Photographic layout: Rod Morris
Printed in Hong Kong

This book provides an introduction to the world of native and introduced New Zealand birds. Each colour spread focuses on a different habitat and the birds that may often be encountered there. The birds depicted are a guide only — many birds, particularly the more common ones, are not confined to just one habitat, and areas differ throughout New Zealand. Different birds may visit different habitats at different times of the year too.

The number of birds depicted in a book of this size is necessarily limited, but I hope that the variety of birds shown will encourage you to try a little bird-watching for yourself. Each species is reproduced close-up and life-sized to provide you with as sharp a focus as possible. A clear look at a bird offers a much better chance of identification. For example, two 'black' birds may appear the same in the distance, but in close-up they may look as different as the blackbird and the starling on page six.

Some of the birds featured are so large that only their head and neck can be illustrated, but that only adds to the challenge of identification. Bird-watchers in the field may only glimpse part of a bird too, but often that is enough to provide identification. Features on the head and neck of a bird may indicate that you are looking at a tui (white throat tufts), a Californian quail (black topknot), a parakeet (red crown), or a silvereye (a tiny circle of white feathers around the eye) — all are good identifying marks.

Beaks can also be particularly helpful to look at. For instance, the colour of the beak of a black swan is distinctive enough, but the shape of a beak can sometimes tell you a whole lot more. A harrier's beak is for tearing flesh, while a white-faced heron's beak is more like a spear, for stabbing things. A house sparrow's beak is thick and strong for crushing grain seeds, whereas a yellowhammer's beak is much finer, perhaps for dealing with finer seed. Once you know what a bird eats, and what habitat it lives in, you are well on the way to becoming a bird-watcher.

We are fortunate to live in what is a 'Land of Birds'. Our native birds, like the kiwi and kea, are extraordinary creatures by world standards, and today bird-watchers travel from all over the world to see them. But when European settlers first arrived in New Zealand, they thought not so much of the remarkable birds in their new land, but of the familiar birds they had left behind — such as the blackbird and the thrush — and which they proceeded to introduce. Self-introduced birds have also steadily arrived here for longer than you might think. Some, like the pukeko, and kahu, the harrier, even have Maori names, yet arrived from Australia in the relatively recent past, after their larger New Zealand counterparts (the flightless moho, and the gigantic Eyles Harrier) faded from the landscape. However, if you glance through any guide to New Zealand birds a pattern emerges that shows most introduced birds as being common, while the majority of native species are rare or uncommon — why is this?

It's almost impossible to comprehend how long New Zealand was free of predatory mammals, but for over 80 million years there were no mice or rats, ferrets or stoats, or cats and dogs here. The primary predators in those times were eagles, harriers and falcons — hunters that relied on swift flight and keen eyes to locate their prey. This is why so many of our native birds nest and feed close to, or on the ground. It's also why so many are flightless, or nearly so. With so much danger from the sky, it was much safer for New Zealand birds to walk. Drab-coloured brown or green plumage also meant that birds wouldn't be as noticeable to an eagle-eyed hunter. Consequently, native birds such as parakeets and weka, saddlebacks and kiwi, are not programmed to recognise that predatory mammals pose a danger to them on the ground.

Today our rarer vulnerable birds are protected on offshore island sanctuaries, such as Tiritiri Matangi Island off Auckland, and Kapiti Island near Wellington. Bird-watchers are always welcome on both of these islands. Such islands preserve this country's unique biodiversity, and the number that are being cleared of rats and other mammalian predators grows every year.

Because endangered native birds like the saddleback are now becoming a little more common, they too have been included in this guide. Other ancient New Zealand birds still to be found on the mainland, as well as in these pages, are the tiny rifleman and the secretive fernbird. We hope this book provides a colourful and informative first step into the rewarding world of bird-watching in a country that, for many New Zealanders, is still a 'Land of Birds'.

Rod Morris

SILVEREYE
(Zosterops lateralis)

Also called waxeye, or white eye, silvereyes are easily distinguished by the distinctive band of tiny white feathers around the eye. Widespread throughout the South-west Pacific, this little bird first introduced itself into New Zealand in the 1850s, possibly blown here on a storm from Tasmania. Silvereyes now range across the country from sea level to the upper forest edge, feeding among the foliage of trees, on nectar, spiders and insects, caterpillars and fruit. Their nest is a beautifully delicate cup of finely woven grass, bound with cobwebs and suspended, hammock-like, from a forked twig.

SONG THRUSH
(Turdus philomelos)

Easily recognised by their creamy-white breast, speckled with brown, and brown upper parts, song thrushes prefer wooded areas, particularly parks and gardens, although they tend not to penetrate too deeply into native forests. They are particularly adept at breaking open the shells of snails, usually by beating them against an 'anvil', such as a rock or some other hard surface, and feed on earthworms and slugs as well. The song thrush is noted for its loud, clear and repetitive song in the spring, usually delivered from a perch high in a tree, or from a rooftop.

HOUSE SPARROW
(Passer domesticus)

House sparrows are abundant in cities, parks and suburban gardens, as well as in rural farmland, particularly where grain is grown or stored, but do not usually occur in unmodified natural environments such as native forests. While male house sparrows have distinctive black bibs, the drabber female may easily be confused with other introduced birds such as finches. House sparrows are 'weavers', building large untidy dome-shaped nests of straw, and both parents share nest duties, rearing up to six young at a time, which leave the nest when about a fortnight old.

STARLING
(Sturnus vulgaris)

Starlings are introduced colonists that may be found in cities and suburbs, and also thrive in open areas, particularly along wild coastal stretches and in remote pastoral habitats. Except when nesting, starlings live in flocks, congregating in large communal roosts at night, and may even travel long distances to roost on offshore islands. Starlings usually feed on the ground, particularly in pasture, on insects and grubs, but will also eat fruit, sometimes damaging orchard crops. Their shiny black plumage is glossed with greens and purple, and in winter each feather is tipped with buff brown, giving a speckled appearance.

9

BLACKBIRD
(Turdus merula)

The male (illustrated) is easily recognised by his black plumage and orange beak. The female and young are dark chocolate brown, with speckling on the breast, and may sometimes be mistaken for the more boldly speckled song thrush. Male blackbirds may be confused with starlings on the ground, however starlings walk and run whereas blackbirds hop. Blackbirds feed on open lawns, searching for earthworms, snails, slugs, and insects, and also eat fruit from orchards and hedgerows in the autumn. Delivered from treetops, the blackbird's song is more mellow than the thrush, and without the thrush's repetitive phrases.

YELLOWHAMMER
(Emberiza citrinella)

Originally from Eurasia, the introduced yellowhammer favours open country, especially rough pasture with hedgerows and scattered scrub. The male has a bright yellow head and breast, with reddish upper parts, while the female is a paler yellow, and streakier. In summer, males sing their distinctive 'chitty-chitty-chitty-sweeeeee' song from some prominent perch — usually along hedgerows and fences. Widespread throughout the country, yellowhammers are typical buntings with a distinctive beak for feeding on fine seeds and insects. They nest in thick vegetation, very close to the ground.

HARRIER
(Circus approximans)

Readily distinguished by their slow, effortless flight, these 'hawks' are well distributed throughout New Zealand. Pastoral development has helped their spread, as has the high number of road-killed possums and rabbits in rural areas. Such 'kills' are a welcome source of additional food, particularly in winter, when young harriers find it difficult to survive. A harrier's plumage may be any shade of brown, from the dark black-brown of a recently fledged juvenile, through mid-browns to the pale greyish-brown of a very old male. Nests are built on the ground, often in swampy places, and the female lays three or four white eggs.

PHEASANT
(Phasianus colchicus)

Pheasants, with their distinctive copper plumage and long tail, are common in rough farming country, especially where overgrown hedges or scrub-filled gullies provide cover. Originally introduced in the 1840s as game birds, they may be found over much of the North Island, and down the eastern coastal strip of the South Island. Pheasants spend much of their time in dense cover, and are often only seen when flushed, when they rocket up and away on noisy wings, cackling in alarm. The nest is situated on the ground in thick vegetation, and may contain six to twelve khaki-coloured eggs.

GREY WARBLER
(Gerygone igata)

The grey warbler is a native bird very much at home on scrubby hillsides in rural areas as well as in forests and suburban gardens. Its sustained trilling song is familiar to most New Zealanders as a harbinger of spring. Throughout the year warblers remain in pairs, searching the outer foliage of trees for insects and caterpillars. The nest is suspended from a branch, and is small and pear-shaped, with a neat access hole woven into the side. Warblers often produce two nests a season but the second clutch is often parasitised by the shining cuckoo.

13

CALIFORNIAN QUAIL
(Callipepla californica)

The male is an easily recognised small bluish-grey game bird, with a black face and topknot, while the female is a more brownish-grey colour, without the black face. Originally from the dry scrublands of California and Mexico, this quail was introduced into New Zealand in 1862, and is now widespread through much of its drier parts, being particularly common in open scrub and mixed farmland. Californian quail live and feed mainly on the ground, in family parties in summer, and larger groups or 'coveys' in winter, and eat mainly weed seeds, but will also take insects, berries and grain.

16

RED-BILLED GULL
(Larus novaehollandiae)

The most common gull in New Zealand, adults are easily recognisable by their bright red bill and legs. Confined to coastal areas on the mainland (except for an inland colony on the shores of Lake Rotorua), they are also widespread on offshore and outlying islands, and are seen in Australia and South Africa. In winter, these gulls disperse from their breeding colonies and may be seen feeding on a variety of food — including anything from fish and marine invertebrates on beaches, worms and insects on flooded pasture, to refuse and scraps at rubbish dumps and sewer outfalls.

BLACK-BACKED GULL
(Larus dominicanus)

More widespread than the red-bill, this is our largest gull — ranging from coastal to alpine areas. Adults have a recognisable black back and a yellow beak with a red spot on the lower tip of the beak. Juveniles have a mottled brown plumage, which slowly changes to the adult's black and white over three to four years. These gulls may nest either in colonies, or as isolated pairs. They lay two to three eggs, usually on cliff ledges or beaches, and because of their general tolerance of people, may even nest on city buildings.

WHITE-FACED HERON
(Ardea novaehollandiae)

Self-introduced from Australia, this heron first began breeding here in the 1940s, and has since become our commonest heron. Its plain bluish-grey plumage, and white face is easily recognisable. Herons stalk their prey, usually in wetlands and along our coasts. They hunt small fish such as bullies, aquatic insects, freshwater crayfish, tadpoles, frogs and lizards. Their nest is usually a loose untidy pile of sticks in a tall tree, where three to five eggs are laid. In winter this heron sometimes gathers to feed in flocks on flooded pasture.

17

VARIABLE OYSTERCATCHER
(Haematopus unicolor)

With an entirely black plumage, and a carrot-red beak, oystercatchers living on beaches at the southern end of New Zealand are easily recognised. However, in northern areas some 'pied' and intermediate 'smudgy' coloured birds also occur! The 'pied' form of this oystercatcher is easily confused with the similar-looking South Island pied oystercatcher (or SIPO) — however, if the line of demarcation between black and white on the chest is a nice clean line it's a SIPO, if it's smudgy it's our variable oystercatcher. Oystercatchers feed on marine invertebrates such as mussels, chitons, polychaete worms and crabs.

BLUE PENGUIN
(Eudyptula minor)

The smallest penguin in the world occurs throughout our coastal waters, breeding along our coast and on offshore islands. It is still common where it is out of the reach of introduced predators such as dogs, ferrets and stoats, even in harbours and urban coastal areas. Coming ashore after dusk, and departing shortly before dawn, little blues may be cautious and secretive on land, but they can also be very noisy — especially during the breeding season — moaning and wailing late into the night. They nest in burrows, or natural crevices, in dense vegetation and occasionally under beachside buildings.

BROWN KIWI
(Apteryx mantelli)

New Zealand's national bird is renowned for its strange, un-birdlike appearance. The flightless kiwi feed mainly on earthworms and insect larvae, which they probe for in the soil, but they will also eat fallen fruit. Kiwi are nocturnal, have very keen hearing, and are difficult to approach in the wild. Unfortunately, kiwi have disappeared from much of New Zealand because their distinctive musky smell enables introduced mammals, especially dogs, ferrets and stoats, to track them to their burrows and kill them. Today they are only safe on offshore islands, and perhaps in a handful of specially protected 'kiwi sanctuaries' on the mainland.

MOREPORK
(Ninox novaeseelandiae)

New Zealand's only native owl, the morepork is common throughout the country in native forest, and has adapted well to smaller bush reserves in urban areas. Morepork are nocturnal and insectivorous, feeding mainly on weta and moths, but they will also hunt small mammals and nestling birds. In urban areas they may be spotted around streetlights after dark, 'hawking' for moths. The bird's characteristic 'morepork' cries can easily be imitated, and with practice it's possible to draw them in for a closer look. By day, morepork commonly roost in dark quiet places in dense vegetation, caves or tree hollows.

ROBIN
(Petroica australis)

With its dark slate-grey plumage, and a creamy-white chest, New Zealand's 'bush' robin's loud cheery song, and friendly, confiding disposition make it a highlight of any forest walk. Often encountered in forest picnic areas, they frequent the forest floor and hop about on long legs, often with an upright stance, as they search the litter for insects, grubs and worms. A quick scratch in the litter, or an overturned scrap of moss or bark will usually draw them closer in search of food — and if you are lucky, may even result in one perching on your leg, or shoe!

SADDLEBACK
(Philesturnus carunculatus)

Named after the chestnut saddle across its back, saddlebacks were once widespread throughout the New Zealand mainland, inhabiting forest and scrub. Today the saddleback is a threatened species, confined to populations on offshore islands — the direct result of active intervention to save it back in the 1960s, when rats threatened their last island homes. At that time, some were carefully transferred to new island homes where they became established, and as a result many island populations exist today. Saddlebacks belong to an ancient family of New Zealand wattlebirds, along with the rare kokako and the extinct huia.

FANTAIL
(Rhipidura fuliginosa)

Instantly recognisable by their flitting, darting flight, and distinctive spreading tail, the fantail's movements are designed to flush insects out of the surrounding foliage so that they may be captured in flight. Fantails are often attracted to humans moving in the bush for the same reason — our movements also disturb their prey. Common throughout New Zealand, in the South Island about one-fifth are a dark sooty black, rather than the more typical pied form. Fantails build a small, neat cup-shaped nest in the bush, often near water, and in a good year three or sometimes even four broods may be reared.

KAKA
(Nestor meridionalis)

Native parrots that prefer undisturbed forested areas, kaka can still be found in certain old-growth forests on the mainland, but numbers are in sharp decline because of predators such as stoats and rats, which raid their nests, killing incubating females and nestlings, and destroying eggs. Fortunately, kaka are still common on stoat-free Stewart Island, as well as on island sanctuaries such as Kapiti and Little Barrier. Kaka feed on fruit, nectar and sap, with their brush-tipped tongues, while their massive beaks are perfect for tearing open rotten wood in search of large wood-boring insects and their fat larvae.

24

Red-crowned parakeet
(Cyanoramphus novaezelandiae)

Once extremely common in the North and South Island, this little green parrot, with its red cap and patch through the eye, is now extremely rare in mainland forests, although it is still common on many offshore islands. Red-crowned parakeets eat a wide variety of plant material, especially seeds, buds, shoots and flowers, as well as nectar and insects. Some foraging takes place high in the forest canopy, but they are equally at home scratching around on the forest floor — which may explain their decline, and that of closely related parakeets with orange or yellow crowns, due to predation.

KERERU
(Hemiphaga novaeseelandiae)

Kereru play a very important role in the ecology of New Zealand forests, as seed-dispersers. Typically they are birds of lowland forests, and forest margins, but today kereru also frequent forested parks and gardens. They are herbivorous feeders, and enjoy the berries, fruits, flowers and leaves of many native and introduced trees and shrubs. Even out of sight, high in the canopy, their loud swooshing flight is unmistakeable. With their small neat heads, metallic green and russet plumage, and proud white chests, these large and very handsome native birds easily compare with their tropical cousins, the fruit doves.

TUI (Prosthemadera novaeseelandiae)

The tui's melodious song, together with that of the bellbird, characterises New Zealand's 'dawn chorus'. With its glossy metallic black plumage and white throat tufts, the tui is one of our handsomest and best-loved native birds. Tui are 'honey-eaters' feeding mainly on nectar from native flowers like kowhai, flax and pohutukawa, but they also eat native insects and fruit. It is a bird of forests, of forested parks and wooded suburban gardens; and in late winter, or early spring, tui range far and wide in search of nectar — especially if there is a kowhai in flower somewhere in the neighbourhood.

PUKEKO
(Porphyrio porphyrio) The pukeko is widespread and common throughout the lowland of New Zealand, inhabiting swamps, marshes, lake edges, as well as adjacent pasture, and even roadside verges, providing there is rough cover nearby. About the size of a domestic hen, pukeko have rich purple-blue underparts, a bright red bill, and a white undertail, which they often flick nervously. Sociable birds, living in permanent groups, they eat mainly seeds, shoots and the roots of aquatic plants, which they hold in one foot as they feed, but will also take insects, frogs, and even the eggs and small chicks of other birds, given the opportunity.

PIED STILT
(Himantopus himantopus) This slim black and white wading bird with its long pink legs is common throughout the North and South Islands of New Zealand, in any wetland where there is sufficient shallow water. Stilts may be seen around swamps, lagoons, estuaries, along riverbeds and out in flooded pasture. They are noisy, gregarious birds that often intersperse their probing for insects and crustaceans in water and mud, with a yapping cry. In autumn, flocks shift to the coast to overwinter.

FERNBIRD
(Bowdleria punctata) This secretive native bird inhabits swamps and rough scrublands, and although not abundant, is still widespread throughout the country. More often heard than seen, fernbird pairs make an 'u-tick' call as they forage in search of insects and spiders. They are weak fliers and prefer to scuttle through the dense undergrowth. About the size of a sparrow with a longish scruffy tail, they can be curious little birds, and if you sit quietly close to where they are calling, they will often approach, creeping mouse-like through the dense scrub, for a clearer look, before disappearing again to forage.

BLACK SWAN
(Cygnus atratus)

The world's only all-black swan (but with white wingtips) is a native of Australia, and was introduced here in the 1860s. Black swans are nomadic, and common throughout the North and South Islands. They prefer relatively shallow lakes and lagoons, where they can feed on submerged aquatic plants, which they reach with their long slender necks. If winter rains cause lake levels to rise, they may shift to flooded pasture. Black swans may either breed in large colonies, or as single pairs.

MALLARD
(Anas platyrhynchos)

Now the most successful of all ducks in New Zealand — adapting quickly to life on ponds in cities, towns and farms — the introduced mallard was first brought out from Europe in 1867, with North American birds added from the 1930s. Since then, mallards have successfully competed with the slightly smaller native grey duck, and now make up 80 percent of dabbling ducks. The male's glossy green head is unmistakeable, but females are often confused with grey ducks. However, the brightly coloured speculum on a mallard's wing is purplish blue, while on the grey duck it is green.

WEKA
(Gallirallus australis)

Weka are characteristically inquisitive, flightless native birds that once inhabited a variety of habitats, from the coast to the high country. Today unfortunately, like so many of our other flightless native birds, their range has diminished through being hunted by introduced mammals, such as ferrets and stoats, as well as dogs. Where you can still find them (and often it is that they find you), weka stalk around, their tails flicking nervously, as they keep a wary eye out for food such as insects or other invertebrates, lizards, the eggs (and young) of other birds, or perhaps some human handout.

KEA
(Nestor notabilis)

Named after its noisy 'kea-a' call, this bird is a native South Islander, where it is confined to the mountainous regions. In the wild they forage for insect larvae, nectar and berries, but these opportunistic parrots are also scavengers and will often fossick around human habitation — particularly tourist lookouts and winter ski resorts — looking for scraps. Unusually for a parrot, kea nest in winter, when much of their habitat is covered in snow, but young birds are slow to develop, and take several months to fledge — by which time it is warm outside again.

RIFLEMAN
(Acanthisitta chloris)

Our tiniest native bird is easily identified because of its size and tailless appearance, and its ceaseless activity searching diligently up and down the trunks and branches of forest trees. But the rifleman is also special because it is the survivor of an ancient group of New Zealand wrens that once scuttled about the trees and forest floor, like so many feathered mice. Of the seven species of wren that we once had, only the rifleman could fly well. The only other survivor, the rock wren, is a poor flier, confined to the mountains of the South Island.